DEVELOPMENT OF

A TIME-SHARED ANALOG COMPUTER

* * * * *

Lorne E. Minogue

and

Cameron G. McIntyre

DEVELOPMENT OF

A TIME-SHARED ANALOG COMPUTER

by

Lorne E. Minogue

Lieutenant Commander, Royal Canadian Navy

and

Cameron G. McIntyre

Lieutenant, Royal Canadian Navy

Submitted in partial fulfillment of
the requirements for the degree of

MASTER OF SCIENCE
IN
ELECTRICAL ENGINEERING

United States Naval Postgraduate School
Monterey, California

1 9 6 0

DEVELOPMENT OF

A TIME-SHARED ANALOG COMPUTER

by

Lorne E. Minogue

and

Cameron G. McIntyre

This work is accepted as fulfilling

the thesis requirements for the degree of

MASTER OF SCIENCE

IN

ELECTRICAL ENGINEERING

from the

United States Naval Postgraduate School

ABSTRACT

The practicability and utility of analog computers, particularly for the solution of applied mathematical problems, is well known. The extension of this usefulness in simple mathematical computations and compensations in the automatic control field is also common knowledge. The purpose of this investigation was to study the feasibility of replacing all the d-c amplifiers by one time-shared d-c amplifier and still retain the versatility of the analog computer. The various functions of addition, multiplication, integration and control applications were investigated. Work was carried out at Advanced Technology Laboratories in Mountain View, California. The authors of this thesis worked under the supervision of Mr. Fred Fitting who originally proposed the idea for study with the intent of investigating broader applications of the DELTASWITCH.

The authors wish to express their appreciation for the assistance and encouragement given them by Mr. Fred Fitting and Assistant Professor R. D. Strum of the United States Naval Postgraduate School in this investigation.

TABLE OF CONTENTS

iii

CHAPTER I

INTRODUCTION

The basic circuit which was to be investigated is outlined in Fig. 1.
It was our intention to study the feasibility and practicability of the
use of such an analog configuration in the controlling of multiple input,
multiple output control processes. It was anticipated that differential
input shown in Fig. 1 would be incorporated, along with its associated
advantages, when computations or compensations of very low signal input
were required. For analysis purposes the simplification of single
input was used and the basic circuit for experimentation and analytical
studies was as illustrated in Fig. 2.

The justification for such a study is warranted by its many applications in both the control and instrumentation field. It was felt that
the time-shared computer would have the following inherent capabilities.

(1) The time-shared computer could replace the sampler, hold and
amplifier in a multiple loop sampled data control system.

(2) It could be used as a time-shared amplifier in any system in
which a large number of operational amplifiers are required. In this
application the multiplying coefficient could be different in each
channel.

(3) If the operations of multiplication and integration could be
justified to comparable accuracy of a continuous analog computer then
the realm of compact, inexpensive computer control would be available
for various types of compensation in the control field.

1

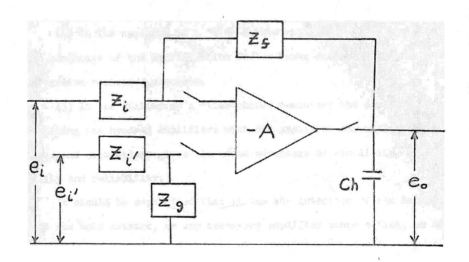

Time-Shared Analog Computer with Differential Input

Figure 1

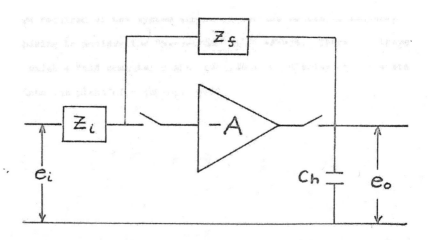

Time—Shared Analog Computer with Single Input

Figure 2

(4) In its capacity as a "time-shared amplifier" there would be the advantage of the amplification factor being determined by the ratio of passive resistive elements.

(5) In its utility as a "time-shared computer" the desirability of replacing one hundred amplifiers with one amplifier for analog computations and simulations gives the added advantage of simplicity, flexibility and reliability.

It should be emphasised that it was the intention of the authors that the hold network, or any necessary amplifier compensation, be simple and inexpensive, in order to make the system competitive and marketable should this feasibility study prove successful. Any deviations from this requirement are outlined as recommendations in the conclusion of this thesis.

It was decided to use the output capacitor in Fig. 2 as the hold network required of the system and to adjust the switching sequence and phasing to achieve the "zero-order hold" effect. Therefore there would exist a held computed signal available for display or for operation into the plant of a control system.

CHAPTER II

THEORY OF OPERATION

2.1 Basic Assumptions:

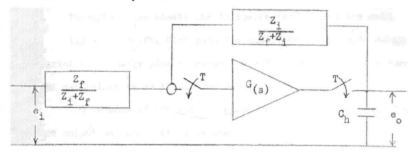

Figure 2A

To analyse the basic circuit of Fig. 2A, it is necessary to sub-
divide the equations to consider the time when the switches are closed
(i.e. normal operational amplifier) and the effect when the switches are
open (i.e. passive computing network circuit) as outlined in Korn and
Korn (1). Because the hold is inherent to and a requirement of the time-
shared computer, and because the hold capacitor does not supply any
energy to the circuit during the computing sequence, the hold capacitor
is not considered when the switches are closed. Therefore, the output
is considered as a combination of a computed voltage level which is
then held at that required computed level. The output on the hold
capacitor is affected to a minor degree by the effect of the input
voltage level when the switches are open. It is the purpose of this
analysis to justify the reduction of the time-shared computer to an

4

equivalent block diagram for the various mathematical operations. This reduction should facilitate studies of problems in the control system field where the time-shared controller may prove practical.

Throughout the thesis the following assumptions are made:

(1) A low drift, high gain d-c amplifier with a wide output voltage variation (usually plus or minus 100 volts) and with a high current output capability is available for $G(s)$.

(2) Because of the switching sequence, and the phasing of the input and output switches, it is assumed that the hold capacitor responds as an ideal impulse integrator which retains the inverted held computed level during the sampling interval.

(3) If the forward gain of the amplifier A is large the relationship between the output voltage e_o, and the grid voltage e_g, are related by

$$e_o = -Ae_g \qquad (2.1)$$

This assumes that the influence of the poles or zeroes introduced by the d-c amplifier have a negligible effect in the analog response.

(4) If the circuit is subjected to a step input then the input and output switches close at the same time as the leading edge of the step input.

5

2.2 Summing Network:

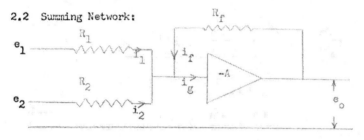

Figure 3

Consider the time when both switches are closed and using Kirchoff's current law as outlined in Korn and Korn (1), the sum of all currents flowing to the grid of the operational amplifier in Fig. 3 are:

$$i_1 + i_2 + i_3 = i_g$$

$$\frac{e_1 - e_g}{R_1} + \frac{e_2 - e_g}{R_2} + \frac{e_0 - e_g}{R_f} = \frac{e_g}{R_g} \qquad (2.2)$$

$$\frac{e_1}{R_1} + \frac{e_2}{R_2} + \frac{e_0}{R_g} - e_g\left(\frac{1}{R_1} + \frac{1}{R_i} + \frac{1}{R_f} + \frac{1}{R_g}\right) = 0 \qquad (2.3)$$

Substitute for e_g from $e_0 = -A e_g$

$$\frac{e_1}{R_1} + \frac{e_2}{R_2} + \frac{e_0}{R_f} + \frac{e_0}{A}\left(\frac{1}{R_1} + \frac{1}{R_2} + \frac{1}{R_f} + \frac{1}{R_g}\right) = 0 \qquad (2.4)$$

Multiply through by R_f

$$e_1 \frac{R_f}{R_1} + e_2 \frac{R_f}{R_2} + e_0\left[1 + \frac{1}{A}\left(\frac{R_f}{R_1} + \frac{R_f}{R_2} + \frac{R_f}{R_g} + 1\right)\right] = 0 \qquad (2.5)$$

6

Therefore

$$e_0 = - \frac{\left[e_1 \frac{R_f}{R_1} + c_2 \frac{R_f}{R_2} \right]}{\left[\frac{A+1}{A} + \frac{1}{A} \left(\frac{R_f}{R_1} + \frac{R_f}{R_2} + \frac{R_f}{R_g} \right) \right]} \qquad (2.6)$$

Define

$$B = \frac{R_f}{R_1} + \frac{R_f}{R_2} + \frac{R_f}{R_g} \qquad (2.7)$$

Provided the forward gain is large and B is small, the numerator of

equation 2.6 reduces to:

$$e_0 = - \left[e_1 \frac{R_f}{R_1} + c_2 \frac{R_f}{R_2} \right] \qquad (2.8)$$

During the time when the switches are open, the output is held at the

computed level in accordance with the hold transfer function Gho(s) as

outlined by Tou(2)

$$Gho_{(s)} = \frac{T_0 s}{1 + T_0 s} \left[\frac{1 - e^{-\frac{t}{T_0}} e^{-Ts}}{s} \right] \qquad (2.9)$$

T_0 is the time constant of the decay of the held voltage dependent

upon the combined voltages through $(R_1 + R_f)$, $(R_2 + R_f)$ and the hold

capacitor C_h acting during the sampling interval T. T_0 also accounts

for the drop off of the held voltage due to loading of the hold

capacitor by subsequent circuitry.

Provided that the sampling interval is short, and provided the

loading of the hold capacitor is negligible, then T_0 can be assumed

to be insignificant. The equivalent transfer function for the hold

capacitor reduces to

$$Gho_{(s)} = \frac{1 - e^{-Ts}}{s} \qquad (2.10)$$

7

Combining equations 2.8 and 2.10, the summed, scaled output
can be represented by the block diagram equivalent Fig. 4.

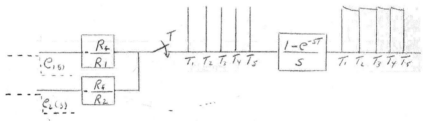

Figure 4

Therefore the overall equivalent expression for the output is

$$e_{0(s)} = \left[- \sum_i \frac{R_f}{R_i} e_{i_{(s)}}^{*} \right] \left[\frac{1 - e^{-sT}}{s} \right] \qquad (2.11)$$

2.3 Multiplication:

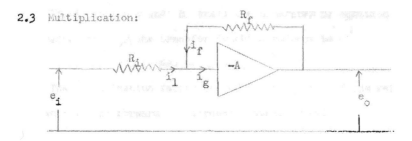

Figure 5

In the analog multiplying circuit of Fig. 5 the following equations from Kirchoff's Law apply

$$i_1 + i_f = i_g$$

$$\frac{e_i - e_g}{R_i} + \frac{e_o - e_g}{R_f} = \frac{e_g}{R_g} \qquad (2.12)$$

$$\frac{e_i}{R_i} + \frac{e_o}{R_f} - e_g\left(\frac{1}{R_i} + \frac{1}{R_f} + \frac{1}{R_g}\right) = 0 \qquad (2.13)$$

Substitute for $e_o = -A_{e_g}$

$$\frac{e_i}{R_i} + \frac{e_o}{R_f} + \frac{e_o}{A}\left(\frac{1}{R_i} + \frac{1}{R_f} + \frac{1}{R_g}\right) = 0 \qquad (2.14)$$

Multiply through by R_f

$$e_i\frac{R_f}{R_i} + e_o + \frac{e_o}{A}\left(\frac{R_f}{R_i} + \frac{R_f}{R_g} + 1\right) = 0 \qquad (2.15)$$

Therefore

$$e_o = -\frac{e_i\dfrac{R_f}{R_i}}{\left[\dfrac{A+1}{A} + \dfrac{1}{A}\left(\dfrac{R_f}{R_i} + \dfrac{R_f}{R_g}\right)\right]} \qquad (2.16)$$

Let B be defined as

$$\left[\frac{R_f}{R_i} + \frac{R_f}{R_g}\right] \qquad (2.17)$$

9

If A is large and B small the numerator of equation 2.16
approaches one and the transfer function reduces to

$$e_o = - e_i \frac{R_f}{R_i} \qquad (2.18)$$

The amplification factor depends on the ratio of the resistive
elements in the forward and feedback paths. Provided the decay of the
held computed level is insignificant then the inverted, multiplied,
held output of a sinusoidal input can be represented in the following
equivalent block diagram Fig. 6.

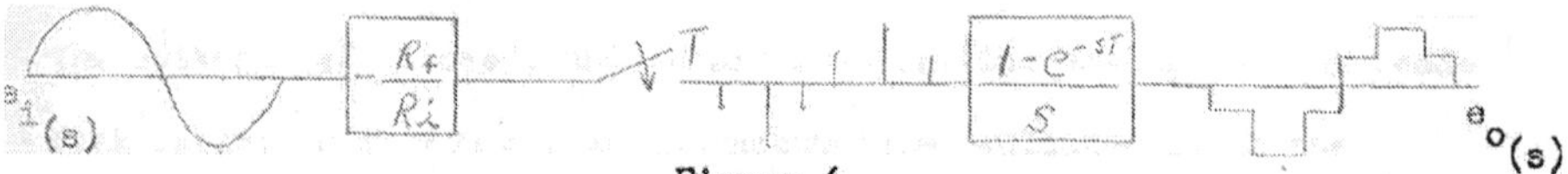

Figure 6

Therefore the overall equivalent output function for multiplication reduces to

$$e_{o(s)} = \left[- \frac{R_f}{R_i} e_{i\,(s)}^{*} \right] \left[\frac{1 - e^{-sT}}{s} \right] \qquad (2.19)$$

10

2.4 Integration:

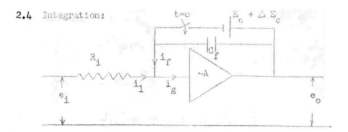

Figure 7

To analyse the function of integration consider the time when
the switches are closed, and consider an initial voltage on the feedback capacitor $E_c + \Delta E_c$, as the accumulated voltage due to the
functioning of the normal operational amplifier (i.e. when switches
are closed) and the operation of the passive computing network
(i.e. when the switches are open). This analog configuration, with
the initial condition $E_c + \Delta E_c$, is illustrated in Fig. 7.

$$\text{For } \tau < 0 \qquad e_{o} = e_g + E_o \qquad\qquad (2.20)$$

Using Kirchoff's first law to sum currents at the grid

$$i_1 + i_f = i_g \qquad\qquad (2.21)$$

$$\frac{e_i - e_g}{R_i} + C_f \frac{d}{d\tau}(e_o - e_g) = \frac{e_g}{R_g} \qquad\qquad (2.22)$$

$$\frac{e_i}{R_i} + C_f \frac{d}{d\tau}e_o - C_f \frac{d}{d\tau}e_g - \frac{e_g}{R_i} - \frac{e_g}{R_g} = 0 \qquad (2.23)$$

Divide through by C_f and substitute the operator p.

$$\frac{e_{i(s)}}{R_i C_f} + p e_o - p e_g - \frac{e_g}{R_i C_f} - \frac{e_g}{R_g C_f} = 0 \quad (2.24)$$

$$\frac{e_{i(s)}}{R_i C_f} + p e_o - e_g\left(p + \frac{1}{R_i C_f} + \frac{1}{R_g C_f}\right) = 0 \quad (2.25)$$

11

Assuming $e_o = - A_{e_g}$

$$\frac{e_i}{R_i C_f} + p e_o + \frac{e_o}{A}\left(p + \frac{1}{R_i C_f} + \frac{1}{R_g C_f}\right) = 0 \quad (2.26)$$

$$\frac{e_i}{R_i C_f} + e_o\left[p + \frac{1}{A}\left(p + \frac{1}{R_i C_f} + \frac{1}{R_g C_f}\right)\right] = 0 \quad (2.27)$$

Let B =

$$\frac{1}{A}\left[p + \frac{1}{R_i C_f} + \frac{1}{R_g C_f}\right] \quad (2.28)$$

as $A \to \infty$ $\quad B \to 1$

$$e_o = - \frac{e_i}{R_i C_f \, p} \quad (2.29)$$

Therefore

$$\int_{e_o = -(E_c + \Delta E_c)}^{e_o} de_o = - \int_0^{\tau} \frac{e_i}{R_i C_f} \, d\tau \quad (2.30)$$

$$e_o + (E + \Delta E_c) = - \int_0^{\tau} \frac{e_i}{R_i C_f} \, d\tau \quad (2.31)$$

$$e_o = - \frac{1}{R_i C_f} \int_0^{\tau} e_i \, d\tau - (E_c + \Delta E_c) \quad (2.32)$$

In integration, if the hold capacitor were not inherent in the time-shared computer circuit and dependent on the switching sequence, it would be necessary to compensate the integrated output by a factor proportional to the pulse width. This would be required because the integrating feedback capacitor is only in the normal operational amplifier integrating mode for the associated period of time. Because of the sampling periods considered, the voltage input e_i, during the time the switches are open, plays an important part in the predicted integrated output.

12

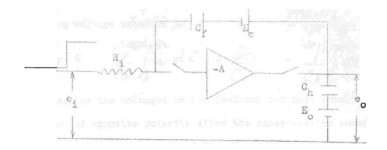

Figure 8

Consider the circuit as shown in Fig. 8 during the time the
switches are open and after a unit step input has been applied. Let
E_o and E_c be the voltages on the hold and feedback capacitors respectively. At the end of each computing period E_o and E_c are equal.
Therefore for the hold period the following passive computing network
is in effect.

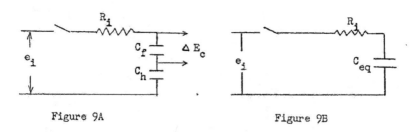

Figure 9A   Figure 9B

. From Fig. 9A, and considering $\dfrac{Q_{c_f}}{C_f}$ and $\dfrac{Q_{c_h}}{C_h}$ as the initial
voltages on the feedback and hold capacitor respectively, the

13

resulting voltage equation is

(2.33)
$$e_{in} = Ri + \frac{1}{C_f}\int_{0}^{\tau} i\, d\tau + \frac{Qc_f}{C_f} + \frac{1}{C_h}\int_{0}^{\tau} i\, d\tau - \frac{Qc_h}{C_h}$$

Assuming the voltages on the feedback and hold capacitor are equal and of opposite polarity after the closed-switch interval, the transfer function of equation 2.33 reduces to

$$\Delta E_c = e_{in}\frac{C_h}{-C_h + C_f}\left[1 - e^{-\frac{t}{RC_{eq}}}\right] \quad (2.34)$$

Where C_{eq} is defined as

$$\frac{C_f\, C_h}{C_f + C_h}$$

Equation 2.34 is the response of the passive computing network to a step input and ΔE_c is a voltage accumulated on the feedback condenser during the time the switches are open. Each time the switches are closed the voltage ΔE_c is transferred to the hold condenser and a discrete stepped voltage output is the result of the integration of a negative step input. The predicted output is illustrated in Fig. 10.

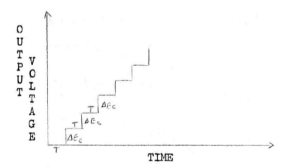

Integrated Output of a Step Input

Figure 10

In order for ΔE_c and t to be linear it is essential that the exponent be in the linear portion of the exponential function. Expand $e^{-\frac{t}{RC_{eq}}}$ from equation 2.34 in a series

$$\Delta F_c = e_{in}\frac{C_h}{C_h+C_f}\left[1-(1-\frac{t}{RC_{eq}}+\frac{t^2}{2(RC_{eq})^2}-\cdots)\right] \quad (2.35)$$

$$\Delta E_c = e_{in}\frac{C_h}{C_h+C_f}\left[\frac{t}{RC_{eq}}-\frac{t^2}{2(RC_{eq})^2}+\cdots\cdots\right] \quad (2.36)$$

Assuming t very small and RC_{eq} large the linear approximation reduces equation 2.36 to

$$\Delta E_c = e_{in}\frac{C_h}{C_h+C_f}\left[\frac{t}{RC_{eq}}\right] \quad (2.37)$$

Substitute for C_{eq}

$$\Delta E_c = e_{in}\frac{C_h}{C_h+C_f}\left[\frac{t(C_h+C_f)}{R}\frac{}{(C_hC_f)}\right] \quad (2.38)$$

Therefore

$$\Delta E_c = e_{in}\frac{t}{RC_f} \quad (2.39)$$

From equation 2.39 and subject to the restrictions of small t and large RC_{eq}, then the scaling characteristics of a continuous analog computer can be achieved by using R and C_f as the scaling factor.

The equivalent block diagram and output function of time-shared integration is given by Fig. 11.

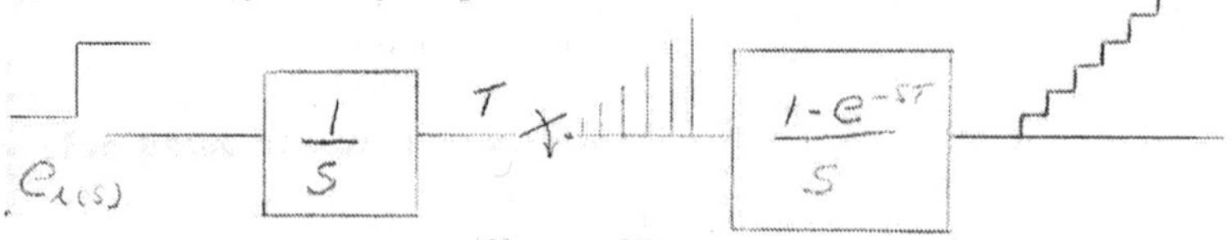

Figure 11

$$e_{o(s)} = \left[e_i\frac{1}{s}\,^*\right]\left[\frac{1-e^{-sT}}{s}\right] \quad (2.40)$$

15

Calculations of ΔE_c

    Case I: Assume $C_f = 1uf$ $t = T = .033$ secs.

$$C_h = 1uf$$

$$R_i = 1 \text{ Meg.}$$

Therefore
$$C_{eq} = \frac{C_h C_f}{C_f + C_h} = .5uf.$$

From equation 2.34
$$\Delta E_c = \frac{e_i}{2}\left[1 - e^{-\frac{.033}{.5}}\right]$$

$$\Delta E_c = e_i \times .0332$$

For exact linear correlation

$$\Delta E_c = e_i \times .0333$$

    Case II: Assume $C_f = 2uf$ $t = T = .033$ secs.

$$C_h = 1uf$$

$$R_i = 1 \text{ Meg.}$$

Therefore
$$C_{eq} = \frac{C_h C_f}{C_h + C_f} = 2/3$$

From equation 2.34
$$\Delta E_c = \frac{e_i}{3}\left[1 - e^{-\frac{.033 \times 3}{2}}\right]$$

$$\Delta E_c = e_i \times .0163$$

For exact linear correlation

$$\Delta E_c = e_i \times .0166$$

    Case III: Assume $C_f = .1uf$ $t = T = .033$ secs.

$$C_h = 1uf$$

$$R_i = 1 \text{ Meg.}$$

16

Therefore

$$C_{eq} = \frac{.1}{1+.1} = .909$$

From equation 2.34

$$\triangle E_c = e_i x .91 \quad 1 - e^{-\frac{.033 \times 1.1}{.1}}$$

$$\triangle E_c = e_i x .277$$

For exact linear correlation

$$\triangle E_c = e_i x .333$$

2.5 Error Analysis:

To determine the errors due to time-sharing of the d-c amplifier
in an analog configuration it is necessary to determine the errors
introduced in a continuous computer and then add any accumulated errors
due to time-sharing. Since only one d-c amplifier is required, it is
assumed that errors due to effects of distortion caused by dynamic
changes of gain, unbalance in the amplifier due to drift, and any other
errors introduced because of amplifier limitations are minimized in the
design of a high gain, high performance amplifier. The design necessary
for such an amplifier is specified in Korn and Korn (1).

The analog operations of addition and multiplication are subject
to error depending upon the value of B given by equation 2.7 and 2.17.
From equation 2.7 the factor $\frac{R_f}{R_1} + \frac{R_f}{R_2} + \frac{R_f}{R_g}$ must be small to justify
that the denominator of equation 2.6 approaches one as "A" approaches
infinity. Therefore this restricts R_g to a very large value and the
practical limits of R_f, R_1 and R_2 specified by Wheeler (3) are 0.1
megohms to 10 megohms. The upper limit of 10 megohms is restricted by
the accuracy of measuring these resistors and the variable resistance
due to temperature effects.

In addition an error is introduced due to the assumption that T_o
of equation 2.9 is insignificant enough to warrant the reduction to
the equivalent transfer function of a perfect zero order hold. This
error is dependent on both the open-switch effects and the loading
effects on the hold capacitor. This error is not critical provided

18

the sampling frequency is 30 cycles per second or faster.

Since addition and multiplication are similar operations the same restrictions on T_o apply. The restriction for B, from equation 2.17, is given by $\frac{R_f}{R_i} + \frac{R_f}{R_g}$. Since high gain is assumed, other factors which determine the size of error, in the approximation that the denominator of equation 2.16 approaches one, is the relative values of R_f, R_i and R_g. These resistors should be restricted to the values previously mentioned.

Integration is accomplished by the open-switch effect of the input on the feedback capacitor and this error is dependent on the linear approximation outlined in equation 2.36. Since integration (during the time the switches are closed) is assumed insignificant, the value of B given in equation 2.28, is of little consequence.

The resultant linear output error of the time-shared integrator is given by the remainder term of equation 2.36. The integration error will be less than or equal to the first term of the remainder.

$$|e| \leq \frac{e_{IN} \, C_h}{R \, (C_h + C_f)} \quad \frac{t^2 (C_h + C_f)^2}{2 R (C_f \, C_h)^2} \tag{2.41}$$

$$|e| \leq \frac{e_{IN} \, t^2 (C_h + C_f)}{2 R^2 (C_f^2 \, C_h)} \tag{2.42}$$

For the percentage error ε it follows that

$$\varepsilon \leq \frac{50 \, t^2 (C_h + C_f)}{R^2 (C_f^2 \, C_h)} \tag{2.43}$$

19

Therefore the percentage error for a given input will be small provided t is very small and for larger values of feedback capacitor. Assume the following typical values

$$C_f = 1uf$$

$$C_h = 1uf$$

$$t = .033 \text{ secs.}$$

$$R = 1 \text{ Meg.}$$

Therefore $\varepsilon \le .1\%$.

An error introduced by sampling in integration is evident in the integral of a constant input to generate a ramp output. Since the integration is not continuous but is accomplished in discrete computed and held steps this error is the equivalent area lost as a result of using a zero order hold. This equivalent loss of area may be determined by $\frac{1}{2} Tx \triangle E_c$ for each sample and so for n samples this error will be

$$\tfrac{1}{2} \, n \, x \, T \, x \, \triangle \, E_c.$$

For a positive ramp output this area will be substractive but for a negative ramp the area will be additive. Provided the sampling period is small and provided $\triangle E_c$ is also small this error will be of little consequence for relatively short integrating times of 10 to 20 seconds.

CHAPTER III

EQUIPMENT REQUIREMENTS

3.1 Amplifier:

The characteristics for a time-shared amplifier used as an analog
computer are essentially the same as those required of a continuous
computer amplifier. A direct-coupled amplifier requires high gain at
low frequencies and this is obtained through conductive coupling
between stages. Because of this type of coupling the output voltage
of a d-c amplifier is sensitive to changes in the d-c reference level
or input voltage for zero input signal.

The output level of a d-c amplifier is sensitive to changes throughout the amplifier and it tends to change or drift with time. This drift
is due to such sources as changes in plate and bias voltage supply,
change in filament voltages, change in resistance values, and changes
in vacuum-tube characteristics. The effects of drift can be reduced
by employing a well regulated power supply for the plate and filament
supplies. Suitable air-conditioning can reduce the effects of changing
resistance due to heating. The amplifier should also have incorporated
a balancing switch which facilitates the balancing without the necessity
for disconnecting the amplifier input and output terminals from the
computer setup.

If a differential amplifier is used then the adverse effects due
to power supply changes, cathode emission changes can be reduced.
Other advantages of using differential amplifiers are outlined in Korn

21

and Kern (1), It was intended that this type of amplifier would be incorporated into the time-shared computer circuit when low signal levels were being multiplied to higher less critical voltage levels. The primary difficulty of including this type of amplifier is the requirement for another switch and therefore the added switching and phasing problems must be considered.

Because the hold capacitor must be charged to the computed level during the switch closure time of 200 microseconds the amplifier must have a high current output capacity.

3.2 Switching Sequences:

Under ideal conditions the opening and closing of the input and output switches occur in synchronism. This requires that the switches close at exactly the same time and open together after the same closure time. In actual practice this action is very difficult to achieve. Therefore it is necessary to investigate the operation of the computer when the closure time of the input and output switches is not identical. Make before break switching causes excessive cross-talk between channels and therefore is undesirable. Only break before make switch action will be considered and employed.

Switching sequences may be considered as one of the following four cases:

. (1) Input switch closes first, opens last.

(2) Input switch closes first, output switch opens last.

(3) Output switch closes first, opens last.

22

(4) Output switch closes first, input switch opens last.

Because the hold network is directly dependent on the operational amplifier output at the end of a computing interval, the amplifier must be disconnected from the hold first. Therefore cases 2 and 3 are not suitable.

In case one, the amplifier is connected into the computer circuit with the feedback path not connected until the output switch closes. This causes saturation of the amplifier and so this sequence is not used.

In case 4, the output switch closes just before the input switch closes and the computer circuit functions until the output switch opens, this leaves the computed voltage on the hold capacitor.

To determine the duration of closure in a switch the circuit shown in Fig. 13 may be used. The phasing of the switch sequence may be checked by observing the duration of closure of the input and output switches on a dual-beam oscilloscope.

CHAPTER IV

EXPERIMENTATION

4.1 Discussion of Equipment:

The Laboratory set-up for confirmation of the time-shared computer
capabilities is illustrated in Fig. 12. As previously outlined two
break before make switches with 65% duty cycle were used in the input
and output of the operational amplifier. A brief description of the
equipment used is as follows.

DELTASWITCH - The Deltaswitch (Appendix A) is a centrifugal pump
primed by a double-lipped scoop which lifts mercury from a sump into
an annular rotating pool in the rotor, from which the mercury is eject-
ed through a small nozzle. After leaving the nozzle, the mercury
impinges upon a series of contact pins in a sequential fashion. After
striking the pins, the mercury falls back into the sump and thus com-
pletes the mercury cycle. The device functions as a switch because
during the time that the mercury stream is in contact with a pin, a
low impedance electrical path exists between the pin and a pole contact inserted into the sump. For experimentation two 100 pin switches,
sampling at 30 c.p.s. were used and phasing was achieved by adjusting
one switch stacked on top of the other and locking the switches when
phasing was checked on an oscilloscope.

OPERATIONAL AMPLIFIER - Because of availability, a PHILBRICK K2-X
d-c amplifier (Appendix B) was used in tandem with a PHILBRICK booster
amplifier K2B1 (Appendix C). This combination gives an open loop gain

24

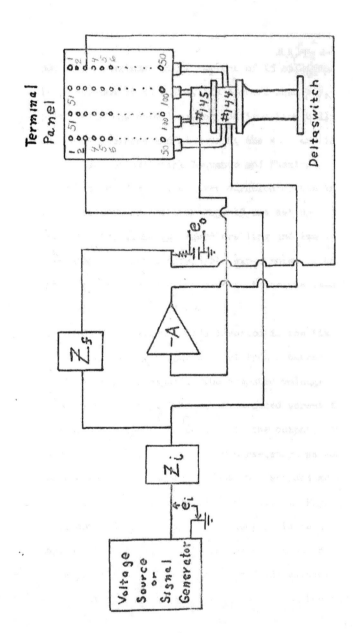

Simple Laboratory Arrangement of a Typical Time-Shared Analog Computer

Figure 12

25

of 30,000 with a maximum output current of 25 milliamperes and an output voltage maximum of 55 volts. To compare accuracy, each experiment was duplicated in the normal analog configuration utilizing the same resistors and capacitors and also using the same amplifier.

4.2 Confirmation of Switching Sequence and Phasing:

In order to confirm the closure duration of the input and output switches the circuit as shown in Fig. 13 was set up and the results are illustrated in Fig. 14. The ragged trailing and leading edges are indicative of the variable duty cycle experienced in this type of switching. The wider of the two closure times was used in the input to the grid of the d-c amplifier.

The appropriate switches were inserted in the time-shared circuit shown in Fig. 12 and the computed level from a battery source voltage of 1.35 volts was photographed. The computed voltage level is shown in Fig. 15. The hold capacitor was then added across the output. Because of sporadic switching operation in the output, the held computed levels were not consistent, so a 68-ohm resistor was added in series with the hold capacitor and the output was recorded across the capacitor. The computed held voltage is illustrated in Fig. 16. The leading edge of the computed pulse in Fig. 14 does indicate a transient due to switching, but the pulse quickly settles out to the required computed level for approximately 150 microseconds. The switching transients are filtered out by the low pass filter action of the hold and are not evident in the held voltage level of Fig. 15.

26

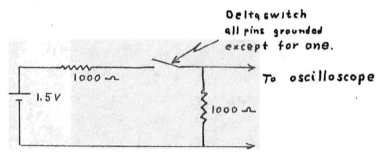

Test Circuit for Closure Time

Figure 13

#145

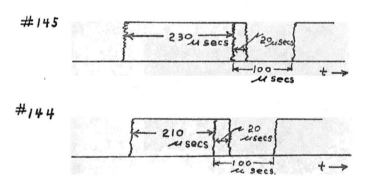

#144

Typical Closure Time Results

Figure 14

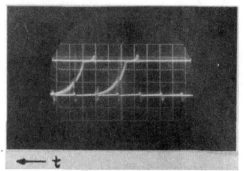

Figure 15

Computed Output Pulse Without Hold

Sensitivites   Voltage - 6.5 volts/cm

              Time   - 0.2 millisec/cm

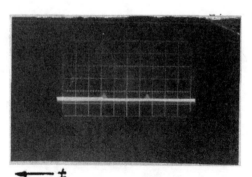

Figure 16

Computed Held Output

Sensitivites   Voltage - 0.5 volts/cm

              Time   - 0.2 millisec/cm

4.3 Addition:

From the theory outlined in Chapter II, the output of two constant voltages with R_f = 1 Meg., R_1 = 1 Meg., R_2 = 1 Meg. should be the inverted held replica of the summation of the two constant voltages. Since this operation necessitates the voltage addition at the node during the computing interval, no problems were anticipated and none were experienced. The results were as predicted and as shown in Fig. 4. Since addition can be considered a special case of multiplication, any restrictions on multiplication would be applicable to addition. Since greater emphasis was placed on multiplying the limitations of addition will be considered under this topic.

4.4 Multiplication:

    4.4(i) Scaling Factors.

From the theoretical analysis of multiplication outlined in Chapter II, it was anticipated that the equivalent block diagram of the time-shared computer would reduce to equation 2.40.

$$e_{o_{(s)}} = \left[-\frac{R_f}{R_i} \; e_{i_{(s)}}^{\;*} \right] \left[\frac{1 - e^{-sT}}{s} \right]$$

To assess the feasibility and accuracy of this reduction, the function of multiplication was considered under the following phases of operation.

    (1) Multiplying factor depending on the ratio $\frac{R_f}{R_i}$.

    (2) Frequency response for a fixed $\frac{R_f}{R_i}$ ratio.

    (3) Successive multiplication on adjacent pins.

The scaling factor or multiplication ratio was varied from a ratio of 1,10,100 to 1,000 and the output was photographed for a constant input frequency of 3 cycles per second. The oscilloscope voltage sensitivities for display of the input and output were adjusted depending on the scaling factor, and the two frequencies were recorded on a dual-beam oscilloscope. The results of these tests are shown in Fig. 17 to 20.

    4.4(ii) Frequency Response.

To illustrate the dependence of the time-shared computer on sampling frequency, and to establish an upper limit on the input frequency, the response of the system was recorded for various input frequencies.

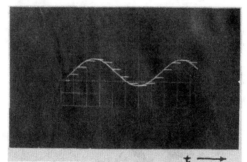

Figure 17

Rf/Ri = 1

Sensitivities Voltage
                  Input 0.5 volts/cm
                  Output 0.5 volts/cm

                Time - 50 millisecs/cm

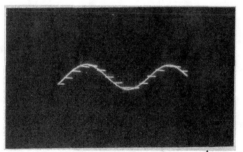

Figure 18

Rf/Ri = 10

Sensitivities Voltage
                  Input 0.5 volts/cm
                  Output 5 volts/cm
                Time - 50 millisecs/cm

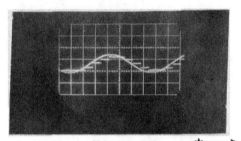

Figure 19   t ——→

Rf/Ri = 100

Sensitivities Voltage
                Input   10 millivolts/cm
                Output 1000 millivolts/cm

                Time - 50 millisecs/cm

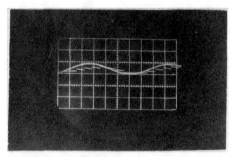

Figure 20   t ——→

Rf/Ri = 1000

Sensitivities Voltage
                Input   10 millivolts/cm
                Output 10 volts/cm

                Time - 50 millisecs/cm

32

The multiplying ratio $\frac{R_f}{R_i}$ was kept constant at one and the input frequency was varied from .5 to 10 cycles per second. The output was photographed and Figs. 21 to 26 illustrate the operation of the zero order hold and the limitations imposed on the system by sampling.

4.4(iii) Consecutive Multiplication.

In order to determine the ability to multiply on two consecutive pins the circuit was set up to allow the input to be multiplied by a factor of two on one pin and then by another factor of two on the adjacent pin. With an input of one cycle per second the output was as shown in Fig. 27.

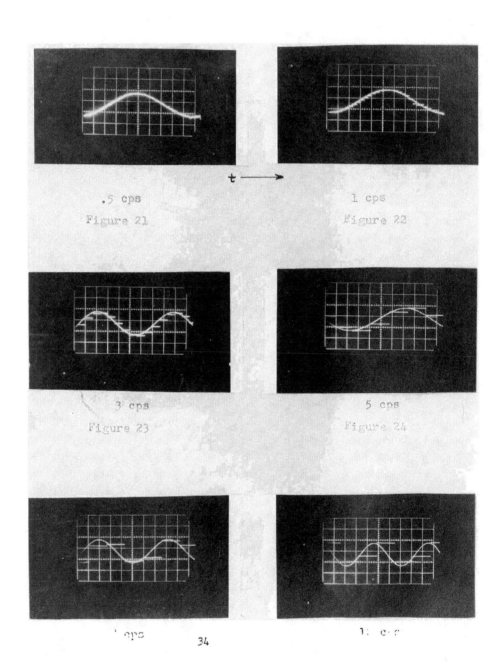

t ⟶

.5 cps
Figure 21

1 cps
Figure 22

3 cps
Figure 23

5 cps
Figure 24

cps

1: e·r

34

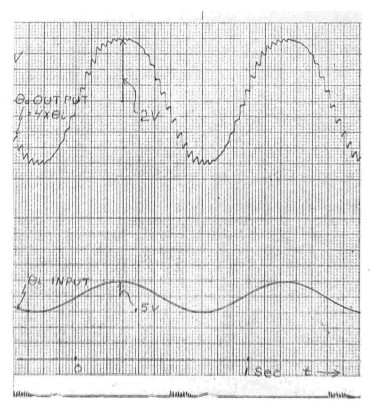

Successive Multiplication

$\theta_0 = 2 \times 2\,\theta_i = 4 \times \theta_i$

Figure 27

4.5 Integration:

   4.5(i) Constant Input.

In Chapter II the analysis of integration indicated that with a capacitor in the feedback path of the time-shared computer the operation of integration could be achieved, but it was dependent on the magnitude of ΔE_c generated during the time the switches are open. To confirm this, the feedback capacitor was connected across the d-c amplifier inside both switches and the circuit was then subjected to a step input. The resultant output voltage indicated that the phenomenon of integration was primarily a function of the passive computing network operating on the input voltage during the time the switches are open.

Because integration errors are a function of gain, the size of resistors, and capacitors, and because these errors are cumulative, it was decided to compare continuous analog integration with time-shared integration. Identical resistors, capacitors and amplifiers were used and an attempt was made to correlate these results. It was hoped comparable accuracy and correlation between the results could be achieved by this technique of equalizing these parameters. Since it was desirable to retain the same scaling capability in the time-shared computer, checks were carried out on integration using various values of feedback capacitor.

Integration was checked for a step input and also for a sinusoidal input. The predicted output to a step input has been outlined in Chapter II.

The values of feedback capacitor in the time-shared configuration of Fig. 12 were varied from 2uf to .luf. The circuit was subjected to a step input and the output for both the continuous and time-shared computers were recorded and are illustrated in Figs. 28 to 30.

4.5(ii) Sinusoidal Input.

The sinusoidal response was recorded for various input frequencies from .1 cycle per second to 3 cycles per second. The values of the parameters were fixed at R_i = 1 Meg., C_f = luf and C_h = luf. The output of both the continuous and time-shared integration are shown in Figs. 31 to 35.

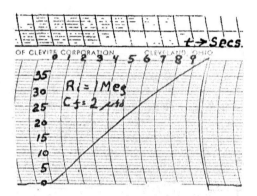

Continuous Integration

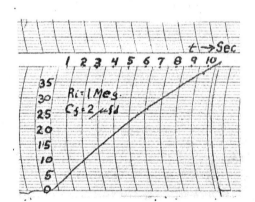

Time-Shared Integration

Figure 28

38

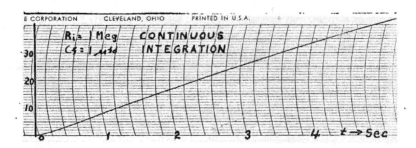

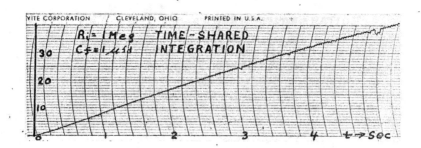

Figure 29

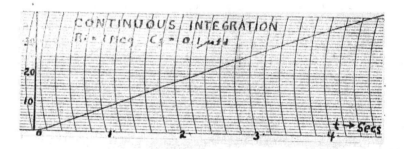

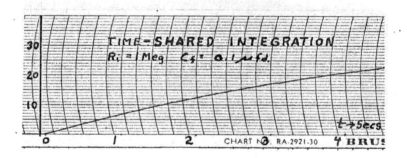

Figure 30

40

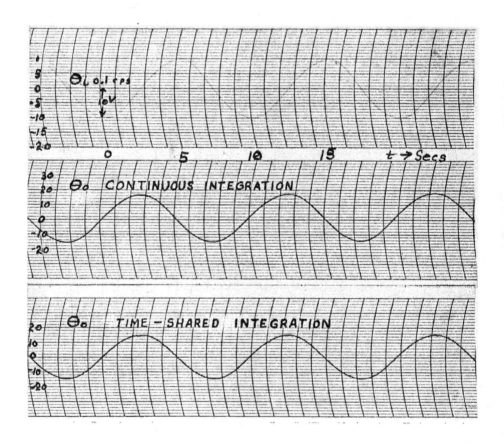

Figure 31

41

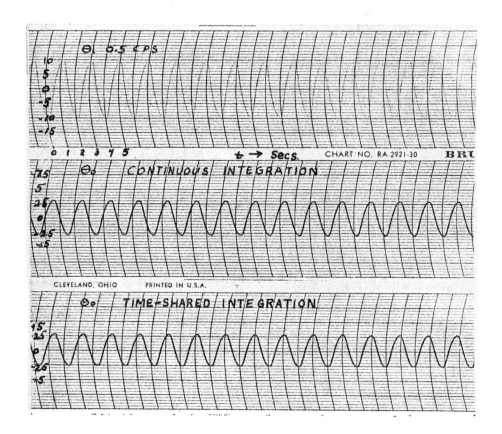

Figure 32.

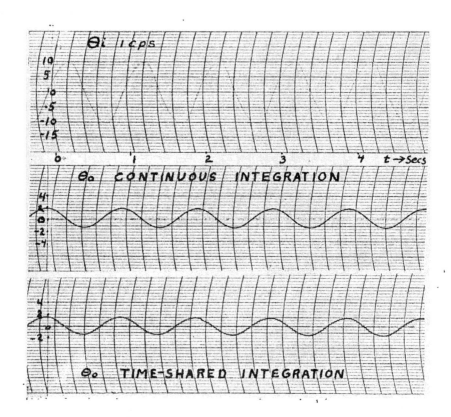

figure 33

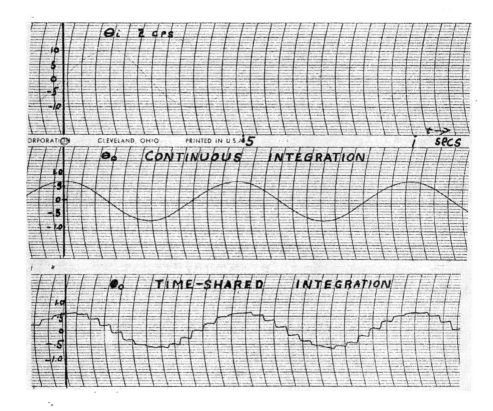

Figure 34

44

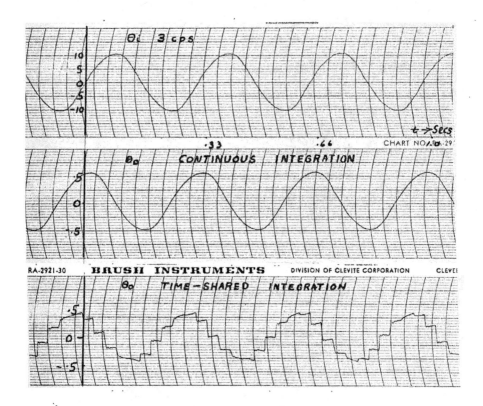

Figure 35

45

4.6 Applications of the Time-Shared Analog Computer:

As typical applications, the summing and integration properties were time-shared in the simulation of a second order control system. The system shown in Fig. 36 is a highly damped, second order, control system with a settling time of about one second to a step input.

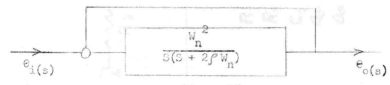

Figure 36

Where W_n= the undamped natural frequency of the system

ρ = the damping factor; chosen to be 0.8

t_s= the settling time $= \dfrac{5}{\rho W_n} = 1$ second

$W_n = \dfrac{5}{\rho T_s} = \dfrac{5}{.8 \times 1} = 6.25$ radians per second

W_c= the frequency of transient oscillations

$= 6.25 \sqrt{1 - (.8)^2} = 3.74$ radians per second.

The response of this system to a unit step input is determined by the following equation

$$O_o = A(1 + \frac{e^{-\rho W_n t}}{\sqrt{1 - \rho^2}} \sin\left(\sqrt{1-\rho^2}\ W_n t - \psi\right) u(t)$$

where $A = 1$ $\rho W_n = 5$ $\psi = \sin^{-1} = \sqrt{1- \rho^2} = 0.6$

$\psi = 180° - 36.5° = 143.5°$

For this system the response to a unit step input is given by

$$\theta_o = 1 + 1.67\ e^{-5t}\ \sin (3.74t - 143.5°)$$

The continuous analog simulation for the system shown in Fig. 36

46

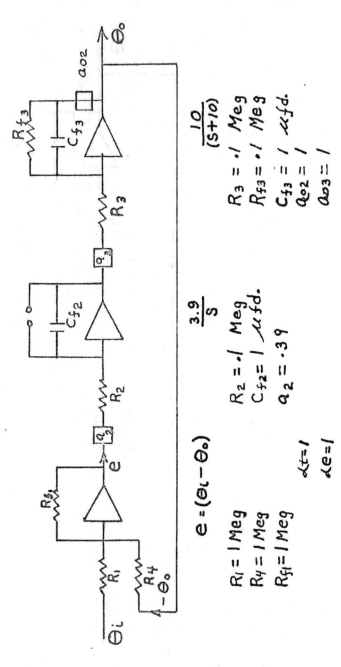

$$e = (\Theta_L - \Theta_o)$$

$$\frac{3.9}{S}$$

$$\frac{10}{(S+10)}$$

$R_1 = 1$ Meg

$R_4 = 1$ Meg

$R_{f1} = 1$ Meg

$R_2 = .1$ Meg

$C_{f2} = 1$ μfd.

$a_2 = .39$

$R_3 = .1$ Meg

$R_{f3} = .1$ Meg

$C_{f3} = 1$ μfd.

$a_{o2} = 1$

$a_{o3} = 1$

$\alpha_T = 1$

$\alpha_E = 1$

CONTINUOUS ANALOG SIMULATION

Figure 39

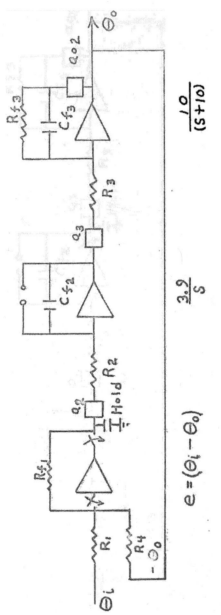

TIME-SHARED SUMMER SIMULATION

Figure 40

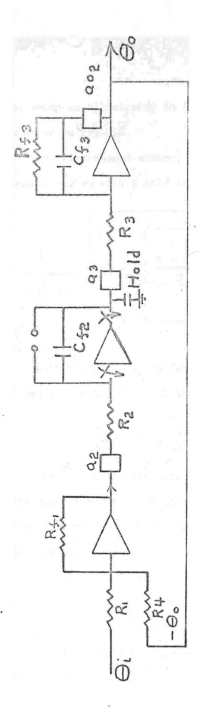

TIME-SHARED INTEGRATION SIMULATION

Figure 41

was setup as illustrated in Fig. 39 and the response to a step input is shown in Fig. 42.

The time-shared summer, or sampled error, response of the same control system with a hold network is outlined in Fig. 37.

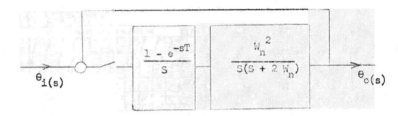

Figure 37

With the same values for W_n and ρ as in the continuous case the closed loop z-transform of the system is given by

$$\frac{\theta_{o(z)}}{\theta_{i(z)}} = \frac{.0199 \ (Z + .882)}{\left[Z-(.8483 + j.112)\right]\left[Z-(.8483 - j.112)\right]}$$

Because of the sampling frequency of 0.033 secs. any calculations were susceptible to inaccuracies due to round off errors and so second order approximations were used as outlined in Tou (2). The results of the calculations were as follows,

M_n – magnitude of peak overshoot – 14%

t_p – time to peak overshoot   – 0.66 secs.

t_s – settling time   – 1.35 secs.

The time-shared analog response to a step input is recorded in Fig. 43.

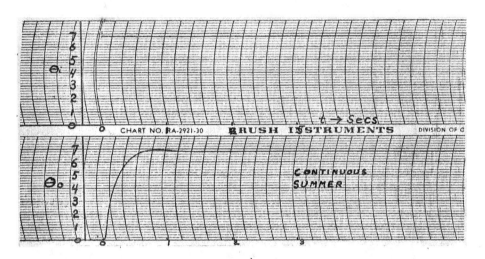

Figure 42

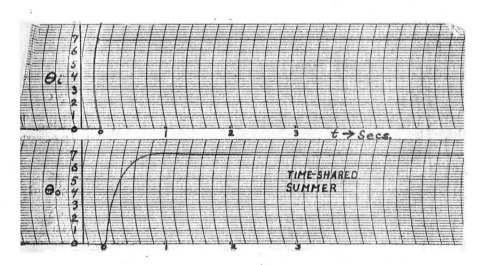

Figure 43

51

The time-shared integration simulation with a hold network is outlined in Fig. 38.

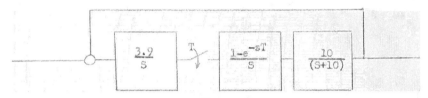

Figure 38

The output of the system was determined by the z-transform analysis and the equation of the output for the closed loop response was found to be

$$\theta_o(z) = \frac{.0363Z}{(Z-1)\left[Z - (.9357+j.179)\right]\left[Z-(.9357-j.179)\right]}$$

Second order approximations were again used and the following results determined,

M_n – magnitude of peak overshoot – 15%

t_p – time to peak overshoot   – 0.56 secs.

t_s – settling time   – 1.27 secs.

The time-shared integration simulation is outlined in Fig. 41 and the time-shared response to a step input is shown in Fig. 45.

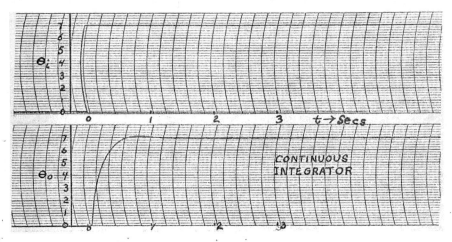

Figure 44

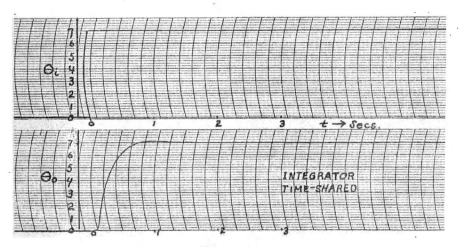

Figure 45

CHAPTER V

SUMMARY OF RESULTS

The basic aim of this thesis was to evaluate the practicability of a time-shared amplifier in an analog computing configuration. From the experimental results outlined in the previous chapter it can be justified that the time-shared computer does retain the elementary computing capabilities of a continuous computer. The time-shared computer is subject to the more stringent limitations imposed by sampling, and also limitations imposed by the experimental equipment available.

From the results of studies of switching sequences and phasing there is indicated the undesirable feature of a transient on the leading edge of the computing pulse. Similarly, the existence of variable switch contact resistance, capacity effects, and variable duty cycle reduce the reliability and accuracy of experimental data. Nevertheless the hold capacitor does function as the low pass filter and zero order hold as predicted. The assumption that the T_o time constant due to loading is insignificant, is only applicable when operating into a high impedance. Loading effects became more restrictive in the investigation of integration.

Although addition was only experimentally verified for the summation of two battery voltages the accuracy was within one percent of the computed voltage level and so the equivalence as a summer, sampler and hold is warranted. Any restrictions on multiplying because of sampling

frequency may be extended equally well to a summation network operating on variable inputs.

The experimental results of multiplication indicate that the equivalent transfer function of equation 2.19, and the assumptions made to achieve this equation, are reasonable. Verification was made of the scaling capability of the time-shared computer as indicated in Figs. 17 to 20. The lack of correlation, especially at the higher ratio of 1,000, may be attributed to the error expected from equation 2.17 and errors introduced by equipment defficiencies and instrument calibration.

Although Shannon's sampling theorem says that in order to recover the input signal, the sampling frequency should be larger than or equal to twice the highest frequency component of the input signal; the results of the time-shared system for a sinusoidal input imply a much higher frequency requirement. A more realistic criterion for the time-shared multiplier would be the requirement that the sampling frequency be at least ten times the highest frequency component of the input. This requirement is evident in the response of the system to a 3 cycle per second input when sampling at a rate of 30 cycles per second. This optimum frequency could have been extended in these experiments to 30 cycles per second by connecting 10 pins in one channel and thereby sample at a rate of 300 cycles per second.

Successive multiplication, as shown in Fig. 27, does not impose any problems of inaccuracies in this type of multiplication due to the loading of the hold in one computing channel by an adjacent

computing operational amplifier. The accuracy of successive multiplication by a factor 2 was found to be very good.

Although the operation of the time-shared integrator can be explained as a combined function of the open and closed switch intervals, the correlation with the larger values of feedback capacitor is quite marked. The lag or error introduced with the .1uf feedback capacitor can be attributed to the non-linearity of the exponential term of equation 2.34. The predicted error from the calculated value of equation 2.34 is 17.4 percent whereas the experimental error is approximately 23 percent. The difference between these errors may be due in part to the loading of the hold capacitor by the recorder. This was found to be more critical with the smaller values of feedback capacitor. In Fig. 29 the sporadic break in the slope of the integrated output was due to lack of consistency in the switching sequence because of instability of the duty cycle of the switches used. The response of the time-shared integrator to a sinusoidal input is in very close correspondence with the output curves of a normal continuous integrator. At the lower input frequencies the time-shared output was recorded at higher recorder speeds and no phase shift, other than the 90 degrees due to normal integration, was apparent.

With regard to the application of the time-shared analog computer, comparisons of the continuous, time-shared calculations, and time-shared experimental results are outlined in Table I.

56

TABLE I

	Time-Shared Calculations		Continuous	Time-Shared Experimental	
	Summer	Integration		Summer	Integration
Mn	14%	15%	2%	2.95%	2.06%
tp	0.66 secs.	0.56 secs.	0.8 secs.	0.95 secs.	1.0 secs.
ts	1.35 secs.	1.27 secs.	1.0 secs.	1.4 secs.	1.3 secs.

It is to be noted that the peak overshoots are much less than the predicted, as determined from second order approximations. The comparison between the settling times are reasonable but the times to peak overshoot show little correlation. These discrepancies may be due to the effects of loading on the hold and also because second order approximations do not consider the transient effects of the zeros in the analysis.

Nevertheless when the two curves, continuous and time-shared, are superimposed the detremental effects of larger overshoot and the faster times to peak overshoot, due to sampling, are not discernable. This would indicate that for the system considered there is little or no difference, particularly for the summer, between the continuous and time-shared simulations.

The simulations that have just been considered are indicative of the capabilities of this basic circuit when employed as a controller. In this particular application, the desirable features of summing, variable gain, and integration are available for utilization or compensation as required by a controller in any one of the one hundred channels that is being time-shared with the one d-c amplifier.

CHAPTER VI

CONCLUSIONS

On the basis of the results of the preceding investigations it can be concluded that the time-shared computer or controller applications are predictable and accurate for the functions of addition, multiplication and integration. Therefore the utility of this concept as outlined in the introduction, wherein its merits as a time-shared amplifier and computer are elaborated, have been justified for further development and application. These results could be improved subject to the following considerations.

The best switching sequence appears to be two break before make Deltaswitches. This type of switching has proven to be the only practical experimental solution to the problem. Nevertheless the Delta-switch introduces the effects of cross-talk, capacitive pick-up, variable contact resistance, variable duty cycle and some switching noise. For signal levels to 10 millivolts these effects do not influence the computing accuracy. Since the switching sequence, described in Section 3.2, and phasing are an inherent requirement of the system other types of switching could be considered in future studies.

(1) A switch manufactured by Magnavox which utilizes a rotating magnetic field instead of the mercury jet should be considered. The details of this switch are given in Control Engineering (4).

(2) To reduce cross-talk and permit definite on-off switching with low noise, the glass enclosed reed relay contact capsules, outlined in

58

Electronics (5), could be employed. The capsules could be actuated by one Deltaswitch at a low impedance level with the phasing accomplished by a capacitor-resistor time constant adjustment on each capsule if required. This type of switching would have the advantage of high impedance level handling of the operational amplifier, and the requirement of only one Deltaswitch.

(3) Electronic switching should be considered but this may be limited by the requirement for a wider closure time in the input than in the output and also the requirement for easy phasing. To accomplish this, with a diode matrix type of switching controlled by two multivibrators for 100 channels, is a major design problem. The complexity and cost of such a dual switching capability may only be justified in large scale, high cost time-shared applications.

Better performance could be expected from a Deltaswitch with a fewer number of pins. This would alleviate the requirement for the small resistor in series with the hold condenser. By using a 64-pin switch, which is available, the sproadic switching operation due to variable duty cycle would be less likely.

Closer correlation between the continuous and time-shared computer could be achieved by sampling more frequently. Under the existing experimental setup the sampling frequency could have been extended to 300 cycles per second by connecting every tenth pin together to give a ten channel capability. It is anticipated that the frequency limit, determined by the sampling frequency, would then be increased by a

factor of ten.

Loading of the hold capacitor was found to be a major factor in
the time-shared circuit. The effects of loading on the hold circuit
were more pronounced in integration than in the other analog operations.
Any attempt to integrate successively did not prove feasible. Because
of this restriction it would be better to have a balanced cathode
follower stage incorporated in the circuit to function as an isolating
stage. This proposed circuit is described in Fig. 46.

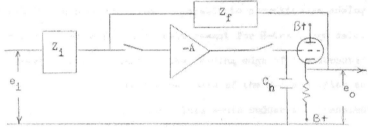

Time-Shared Analog Computer with Cathode Follower Hold

Figure 46

Because the cathode follower would isolate the hold capacitor
from the open switch effects of the input voltage, and because the
cathode follower functions as the energy source, the time constant T_o
would be insignificant and a more perfect zero order hold would result.
Similarly the computation of ΔE_c in equation 2.34 would reduce to

$$\Delta E_c = e_{IN} \left[1 - e^{-\frac{t}{RC_t}} \right] \qquad (6.1)$$

This simplification would retain the scaling capability outlined
in integration. If the same requirements for linearity are imposed on

the exponential $e^{-\frac{t}{RC_f}}$ then integration should function as described by equation 6.1 and greater accuracy would result over a wider range of values of feedback capacitor.

The block diagram equivalent of the time-shared computer has been justified by experimental results in the various phases studied. It is appreciated that the use of such equivalence would be cumbersome in setting up simple problems or compensations. Provided the sampling frequency is increased, by methods previously described, the time-shared computer may then be considered comparable to a continuous analog system and therefore there would be no requirement for Z-transform calculations.

To prevent the transient on the leading edge of the computed pulse it is possible to compensate the response of the d-c amplifier so the amplifier has a slower rise time. This would reduce the transient effect even though this has not been a factor in the computing accuracy of the circuit. The low frequency gain of the amplifier should be as high as possible to minimize error and to increase the performance of the circuit at low input voltage levels and at faster sampling frequencies.

CHAPTER VII

SUGGESTIONS FOR FUTURE ANALYSIS

A study should be made of the feasibility and practicability of the circuit outlined in Fig. 46.

Investigations should be carried out to determine the upper frequency limit of sampling. This upper limit would be a function of the open loop gain of the amplifier, the duration of closure time of the switches, and particularly the generation of $\triangle E_c$ during the time the switches are open.

Successive integration at the higher sampling frequencies should be investigated using the circuit outlined in Fig. 46. The ultimate criterion for correlation between the time-shared and continuous computer would be the comparable solution of a simple second order differential equation.

Studies could be made of the types of switching recommended in the conclusions.

Investigations could be made of the function generating capabilities of the circuit in Fig. 47. The function generating characteristics were checked for the basic circuit outlined in Fig. 2. The circuit did function and the output was comparable to that expected from a single pole lag network. Time precluded further investigation of this time-shared function generating phenomenon. If the balanced cathode follower, suggested in Fig. 46, were incorporated, the mathematical analysis is much simpler and does indicate this desirable feature is

62

available.

In this analysis it was assumed that the passive computing network was of prime importance in function generation. Similarly, it
was assumed that at the end of each computing interval, the voltage
across the output of the cathode follower and the voltage on the feedback capacitor were equal.

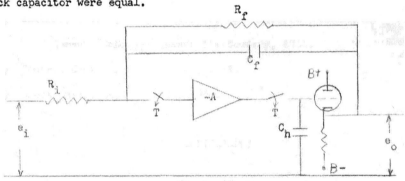

Figure 47

From a similar analysis, as outlined for integration, it was determined
that the controlling transfer function for the time-shared configuration of Fig. 47 would be given by

$$e_o = \frac{e_i}{R_i R_f C_f \left(S + \frac{R_i + R_f}{R_i R_f C_f} \right)} \qquad (7.1)$$

From equation (7.1) the output would be stepped held exponential
curve controlled by the pole of the ratio of $\frac{R_i + R_f}{R_i R_f C_f}$ and attenuated
to a level determined by the ratio of $\frac{1}{R_i R_f C_f}$.

63

REFERENCES

(1) Korn and Korn, Electronic Analog Computers,

McGraw-Hill Book Company, 1952.

(2) Tou, Julius T., Digital and Sampled-Data Control Systems,

McGraw-Hill Book Company, 1959.

(3) Wheeler, R.C.H., Basic Theory of the Electronic Analog

Computer, Donner Scientific Company, 1958.

(4) Control Engineering, November 1958.

(5) Electronics, September 30, 1960.

BIBLIOGRAPHY

Wass, C.A.A., Introduction to Electronic Analog Computers
McGraw-Hill Book Company, New York, 1956.

Ragazzini, J. R. and G. F. Franklin, Sampled Data Control Systems
McGraw-Hill Book Company, New York, 1958.

Jury, E.I.J., Sampled Data Control Systems,
John Wiley & Sons, Inc., New York, 1958.

West, J. C., Servomechanisms, English Universities Press, LTD.,
Saint Paul's House, Warwich Square, London, 1953.

Truxal, J. G. Control Engineers Handbook,
McGraw-Hill Book Company, Toronto, 1958.

Deltaswitch*

DELTASWITCH is a unique commutating device utilizing a mercury stream for a wiper arm. A rotor lifts mercury from a pool; centrifugal acceleration forces it through a nozzle as a high velocity stream. This rotating mercury jet contacts peripheral commutator contacts, forming a very low electrical resistance path between it and a common contact in the mercury pool. DELTASWITCH has no brushes; therefore the problems of high noise, contact bounce, arcing, mechanical wear and duty cycle changes are completely avoided.

*Patent 2,782,273

Deltaswitch is a custom-engineered unit. General specifications are shown below and on the reverse side.

SPEED RANGE — Optimum: 1800 to 3600 RPM. Special models: to 6000 RPM.

NUMBER OF CONTACTS — Any number of commutator contacts to 120.

NUMBER OF POLES — Deltaswitch is a single-pole commutator. Multipole operation is achieved by stacking units and driving them with a common motor.

POLE PHASING — Phase relationship is adjustable. Once set, phase angle remains constant within ±15 minutes of arc over long periods of time.

DUTY CYCLE — Duty cycle (ratio of time in contact to time between initiation of successive pulses) can be varied over a wide range by factory adjustment. Standard models provide a duty cycle of 50% with an accuracy of ±7%. Low duty cycles (under 40%) with a large number of contacts are not recommended.

CONTACT RESISTANCE — Less than 1 ohm (with a 1 volt potential difference).

INSULATION RESISTANCE — Approximately 120 megohms. With Deltaswitch operating, 150 volts between adjacent contacts of a 64-contact switch results in a leakage current of 1 to 1¼ microamperes.

NOISE LEVEL — Primarily a function of switch speed and input resistance. Approximately 10 microvolts at 3600 RPM and 100 ohms.

GENERAL CHARACTERISTICS
Design Center Electrical Characteristics

Gain:
30,000 dc open loop (depending upon the application — see text)

Response: — Small signal:
1 μsec rise time with band width over 250 kc when used as a unity- gain inverter

Drift Rate:
±5 mv per day referred to the input (see text — "DRIFT")

Differential Input Levels:
Impedance: — Either input:
Above 100M (open grid)

Voltage Range:
Inputs together (common mode): −50 to +50 volts

Current: — Either input:
Typically less than 10^{-8} amp (insulation leakage)

Bias Required for Balance:
Adjustable from 0.8 to 1.8 volt between pins 1 and 2 (pin 1 positive with respect to pin 2) (See figures 2, 3.)

Output Capabilities:

Output Voltage	Output Current (steady state)	
	Normal	Case HP*
−100v	−1.0 ma	−3.0 ma
0v	±3.2 ma	+0.6 ma
		−6.0 ma
+100v	+6.0 ma	+3.0 ma

Maximum transient output current is very much larger in the positive direction, but is the same as above in the negative direction.

*With a 100K, 2-watt resistor connected between pin 6 (output) and pin 3 (−300 vdc). This necessarily operates the 6AN8 at its maximum plate dissipation and shortens tube life.

Physical Characteristics

Power Required: (for full output)
Normal Operation: (100K load)
11.8 ma at +300 vdc
8.2 ma at −300 vdc
0.75 amp at 6.3 vac or vdc

Case HP
17.8 ma at +300 vdc
12.2 ma at −300 vdc
0.75 amp at 6.3 vac or vdc

Tube Complement:
1 12AX7 or 7025
1 6AN8

Casing:
Molded plastic, sealed unit

Temperature:
Maximum allowable case temperature (hot spot)
−65°C (149°F)
(see text)

Base:
Octal plug

Dimensions:
Overall: 4½ in. h
Above Socket: 1⅞₆ in.
× 2⅛ lg. × 4¾₂ in. h

Weight:
Installed: 3.0 oz.
Packed: 5.6 oz.

GENERAL DESCRIPTION

The Model K2-X is a high gain, wide band, plug-in, dc operational amplifier, designed and constructed for use as a basic subassembly in analog computer and instrument applications. It is primarily useful in feedback circuits where a high open loop gain and an output voltage range of from minus to plus 100 volts are required. The open loop dc gain for normal operation with a ±60 volt swing and a 50K load is 30,000. With a ±100 volt swing, the dc gain may decrease to 5,000. With these units, computing devices of all speeds can be assembled with a minimum of external circuitry. A schematic diagram of the K2-X is shown in figure 4.

The K2-X will perform the same operations as the K2-W but needs much better ventilation. It features balanced differential inputs for low drift, high input impedance, low output impedance, high performance, and economy of operation. Its range of operation is from dc to above 250 kc.

With appropriate circuitry, the K2-X maintains the two inputs at nearly equal potentials. The residual offset can readily be biased out. (See BIASING METHODS.) Operationally, the K2-X plugs into the same socket as the K2-W, and uses the same connections for power and for computing signals. Because of its greater band width, the K2-X will not tolerate quite as large a direct capaci-

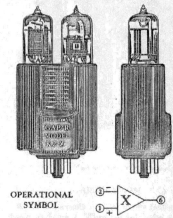

OPERATIONAL SYMBOL

Figure 1. K2-X Operational Amplifier

tive load as does the K2-W. Also, its output must not be grounded.

APPLICATIONS

The K2-X Operational Amplifier can be used for analog computation in feedback systems of any complexity. It is entirely compatible with the K2-W and the two can be used in the same assemblages, each being used to exploit its own special characteristics. The K2-X permits steeper wave fronts and greater signal excursions. Also, its greater output power allows the use of computing networks that require higher voltages and currents than are possible with the K2-W. However, be sure to provide ample ventilation.

Figure 2. Extra Wide Range Amplifier

For straight amplification, the two amplifier tandem arrangement shown in figure 2 is very useful. It offers a more than tenfold speeding-up of the response time for higher gains. Separate resistors can be used for each of the feedforward and feedback elements with binary or decade switching, or the two controls can be ganged to provide a single gain adjustment of extraordinary dynamic range.

With this circuit, (See figure 2.) the effective input noise has been measured to be very much less than one millivolt.

GENERAL SPECIFICATIONS

Output:
Signal:
    Maximum: ± 25 ma at 0 to ± 55 vdc
Load Impedance:
    Minimum for 50 volt output: 1.7K
    Maximum capacitance recommended
    between terminal 6 and ground: 50 μf
Impedance:
Open loop:
    Approximately 250 ohm,
    looking back into terminal 6.
Gain:
    Approximately 0.8
Input:
Signal:
    ± 70 vdc maximum
Impedance:
    Approximately 1 meg
    shunted by 10 μf
Power:
    25 ma at $+300$ vdc quiescent
    35 ma at -300 vdc quiescent
    0.9 amp at 6.3 vac 50-60 cps
Thermal Dissipation:
    24 watts
Tube Complement:
    Two 6CM6
Casing:
    Molded plastic; sealed unit
Base:
    Octal plug
Height:
    5 in. overall
Weight:
    Installed: 3 oz.   Packed: 5½ oz.

GENERAL DESCRIPTION

Model K2-B1 Booster Amplifier is a follower designed for use with Model K2-W or K2-X Amplifiers to increase their available output current. Output currents as high as ± 25 milliamperes at ± 55 volts can be obtained from the unit.

The Model K2-B1 is conservatively designed to avoid self-destruction under severe overload and accidental grounding conditions. It is enclosed in the same type case as the K2-W, but its dissipation is almost five times that of the K2-W. It can be operated in a well ventilated HK Operational Manifold. If it is to be enclosed, provision must be made for getting rid of its 24 watts dissipation without raising the case's hot-spot temperature above 70°C (158°F).

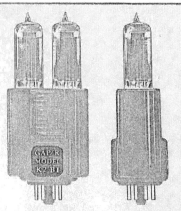

Figure 1.   Model K2-B1, Booster Follower

APPLICATION

K2-B1 is normally used inside the feedback loop as shown in figure 2. When applied under these conditions, the usual operational relationships obtain; that is, $e/e_1 = -z_f/z_1$. The output of the K2-W or K2-X, because of the high loop gain, will automatically adjust itself to enforce this condition; that is, within its ratings it will compensate for attenuation and offset in K2-B1 and reduce its output impedance.

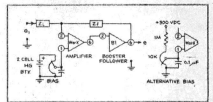

Figure 2.   Typical Application
of a K2-B1 to Operational Circuitry

Figure 3 shows the installation of a K2-B1 in an operational manifold, where it is plugged in adjacent to a K2-W. The K2-B1 can be used in most of the applications described in the K2 Applications Manual (available upon request).